Digitalising Water – Sharing Singapore's Experience

Digitalising Water – Sharing Singapore's Experience

NG Joo Hee
Chief Executive

Harry SEAH
Deputy Chief Executive (Operations)

PANG Chee Meng
Chief Engineering & Technology Officer

PUB, Singapore's National Water Agency

Published by   **IWA Publishing**
Republic – Export Building, 1st Floor
2 Clove Crescent
London E14 2BE, UK
Telephone: +44 (0)20 7654 5500
Fax: +44 (0)20 7654 5555
Email: publications@iwap.co.uk
Web: www.iwapublishing.com

First published 2020
© 2020 IWA Publishing

British Library Cataloguing in Publication Data
A CIP catalogue record for this book is available from the British Library

ISBN: 9781789061864 (Paperback)
ISBN: 9781789061871 (eBook)

This eBook was made Open Access in November 2020.

Contents

Foreword

Over the last decade, almost every aspect of our lives has become digital – from how we deal with finances to the ways in which we are entertained. Pervasive availability of powerful and interconnected technologies, including the Cloud and AI, are leading many businesses and governmental organisations on a journey of digital transformation.

While new technologies have disrupted traditional business models, they have also created immense opportunities for new products, services and new business models that are more sustainable and cost-effective. These changes are set to impact almost every sector, including water.

The IWA Publishing Digital Water Book Series collates perspectives, experiences and best practices to enable its readers to develop a strategic response to the possibilities offered by new digital technologies.

This first book in the series is compiled by Singapore's National Water Agency, PUB. PUB is an internationally renowned, award-winning statutory board that manages Singapore's water supply, water catchments and used water in an integrated way. The holistic approach adopted by PUB has resulted in a lower dependence on external water sources through diversification, including through desalination, stormwater storage and high-quality reused water.

Digitalising Water – Sharing Singapore's Experience outlines PUB's vision and provides examples of some of its digital initiatives. Here, digital becomes an integral vector in an integrated water management approach.

© IWA Publishing 2020. Digitalising Water – Sharing Singapore's Experience
Author: PUB, Singapore's National Water Agency
doi: 10.2166/9781789061871_vii

As digital disruption continues to reshape business domains, organisations will be under pressure to adapt to change. Doing so will allow them to take full advantage of the opportunities ahead. The IWA Publishing Digital Water Book Series will, hopefully, offer your organisation some perspectives on the effective execution of digital strategies for water management.

It is meant to support you as you become more proactive in the digital domain, help you turn digital threats into opportunities, and allow you to leverage digital to create competitive advantages and to enhance performance.

The book series will not be limited to addressing new technologies or systems. The most successful digital transformations depend equally on the sound strategies for management of processes and people. This will be a focal point in future volumes.

We hope you glean useful insights from this series, and ideas that help you along in your own digital journeys.

Professor Vladan Babovic
National University of Singapore
Editor-in-Chief, IWA Publishing Digital Water Book Series

Preface

Water – specifically, the lack of it – occupies a special place in the Singaporean consciousness. We simply do not have the space to collect and to store enough of the rainwater that we will need for later consumption. Water scarcity compels us to turn to recycled wastewater and even dearer desalinated seawater in order to quench new thirst. Indeed, the enduring sustainability and security of Singapore's water supply present, no less, an existential challenge for the country.

Smart Water

Imagine driving a car which has neither speedometer nor fuel gauge. This is essentially analogous to what water end-users find themselves in today. The water meter outside every Singaporean home is the dumbest, albeit highly accurate, of sensors. In order to keep track of water consumption, the consumer needs to physically sight the meter dial, make a record of the counter reading, and then return and repeat the same at a later chosen time. As physical verifications of meter readings are cumbersome and costly, they are done infrequently. For billing purposes, water meters in Singapore are read every other month. Any more often would make it uneconomical.

The challenge for PUB Singapore is to give our customers the water equivalent of the speedometer and fuel gauge, and so empower them to become smarter users of water. Battery and wireless technology, together with cloud computing, have improved to such an extent that a computer-enabled water meter now costs no more than the brass block of old. Yet, such a smart sensor that can measure, register and wirelessly transmit usage data to

© IWA Publishing 2020. Digitalising Water – Sharing Singapore's Experience
Author: PUB, Singapore's National Water Agency
doi: 10.2166/9781789061871_ix

remote servers, which in turn calculate and analyse before proceeding to inform the consumer almost instantaneously, promises to change the game entirely.

Our tryouts of smart water meters show that those who receive timely usage information are able to meaningfully adjust behaviour and become more efficient users, saving water and money in the process. These experiments have convinced us that the digitalization and constant availability of water use information will help to change consumer behaviour, leading to significantly reduced consumption. They also confirm that that our broader ambition to digitalize Singapore's entire water system is a worthy and worthwhile one.

The Future is Digital

There can be no denying that future of water is digital. If machines can do something faster and cheaper, and perfectly all the time, why deploy an expensive human – who invariably gets tired and is prone to mistakes – to the job? If computing power can help us make better decisions, why leave anything to chance or human foibles? Remote sensors of all kinds – which are now inexpensive, reliable, and work round-the-clock – can make real-time system-wide surveillance, something manually impossible, imminently achievable. The digital future beckons!

Hence, the entirety of PUB Singapore is on a grand pivot, away from the slow and burdensome ways of the past towards a future that is smart, speedy and precise. Many systems are getting built, many processes have been re-engineered, and many jobs have, is and will be transformed. This is happening across the length and breadth of our organization, in our plants and in the field, on the frontlines and in the backroom.

Indeed, there is no other way around this, and we know we risk becoming redundant should we opt to shy away from the effort. And so, we have chosen to put digitalization at the centre of PUB's transformation. For if we do not learn a new set of *kungfu*, we will be unprepared for, and be beaten by the future.

Peter Joo Hee Ng
Chief Executive
PUB, Singapore's National Water Agency

Chapter 1

Introduction

Singapore's water journey is at a crossroads.

For decades, the city-state's national water agency PUB has diversified and expanded its sources of water to ensure a reliable supply for residents and businesses.

In the 1970s, the nation began a massive effort to clean up its rivers and waterways and increase its catchment area to collect rainwater, adding a new source of water to its imported water from Malaysia. The country's water catchment area now spans two-thirds of its total land area, including 17 reservoirs island-wide to store rainwater.

When the technology to reclaim used water improved and became more affordable in the 1990s, PUB set up a team to explore its potential application in Singapore. After the utility completed extensive tests and audits and found that it could produce high-grade, reclaimed water in a safe and sustainable way, NEWater was conceived and introduced to the public in 2003. Just two years later, in 2005, PUB added a fourth national tap with the opening of Singapore's first desalination plant.

The creation of the Four National Taps, namely water from local catchments, imported water, NEWater and desalinated water, has cushioned Singapore's water supply against dry weather and reduced its water vulnerability. Along the way, PUB has also inculcated water conservation as a way of life,

© IWA Publishing 2020. Digitalising Water – Sharing Singapore's Experience
Author: PUB, Singapore's National Water Agency
doi: 10.2166/9781789061871_0001

decreasing the average water use per person per day from 165 litres in 2000 to 141 litres in 2019.

Singapore's innovation and investment in its water sector has made it an internationally acclaimed city for integrated water management, as well as an emerging global hydrohub. It is a leading centre for business opportunities and expertise in water technologies, with a vibrant and thriving ecosystem of about 180 water companies and more than 20 water research centres spanning the entire water value chain.

As with many other nations now, however, it faces a confluence of water challenges in the coming years.

Singapore's water demand is slated to nearly double by 2060, from its 430 million gallons per day currently. The country is also bracing for climate change, which has already led to more intense rainfall and longer dry spells, impacting rainwater collection and flood prevention efforts, as well as rising sea levels.

Digitalisation can help alleviate these and other issues. It enables water utilities to collect vast amounts of data, analyse it and produce meaningful insights for more informed and better decision-making. Utilities can also automate processes for efficiency and reap other benefits.

PUB Chief Executive Officer Ng Joo Hee has said: "Digitalisation gives us a huge opportunity to advance our mission in previously unimaginable ways. It is absolutely a gamechanger."

A DIGITAL ROADMAP FOR THE FUTURE

In 2018, PUB unveiled its SMART PUB Roadmap to digitalise Singapore's entire water system. This will optimise the utility's operational capabilities to meet the nation's future needs.

The roadmap outlines PUB's key projects in four major areas: smarter water quality management with artificial intelligence and automation; key network improvements with predictive intelligence; integrated customer engagement with water usage data; and smarter work redesign with automation and robotics.

In smarter water quality management, the utility has started pilot trials on technologies such as a fleet of cost-efficient and highly versatile robots that can continuously monitor the water in Singapore's reservoirs.

In key network improvements, PUB has assembled a hydrometric network with water level sensors and flow sensors, together with a constellation of

Closed Circuit Television cameras (CCTVs) along waterways and throughout Singapore.

These, together with other sensors have vastly expanded Singapore's ability to monitor the condition of its water network. Prior to these digitalisation efforts, manual checks could only reveal the condition of the network at specific points.

Now, PUB can oversee the network in real-time and in minute detail. With digitalisation, the utility can further reduce leaks and non-revenue water, both of which are perennial problems in the water sector.

PUB is also keen to empower customers with near real-time information on their water consumption to encourage better water-saving habits and encourage behavioural change towards conserving water. A previous smart trial with smart water meters installed at about 800 households achieved average water savings of up to 5%, due to early leak detection and adoption of water-saving habits which was largely driven by the in-app gamifications. By 2023, PUB will install smart water meters across seven locations in Singapore as part of the first phase of the Smart Water Meter Programme.

With automation and robotics, utilities can also redesign work processes to make them more efficient. PUB is testing out automated lab analysers, which can test two to three times more samples than human operators in the same amount of time, and allow analyses to be carried out around the clock. It is also exploring virtual reality training for operators, and other digitalisation systems.

These and other digitalisation projects will help PUB to become more efficient and productive, preparing itself to be a smart utility of the future.

THE SINGAPORE DIGITAL WATER STORY

Despite its dividends, digitalisation is not always easy or straightforward to implement. The choices in systems and technologies can be overwhelming, and the upfront investment needed may be daunting, particularly for utilities in smaller municipalities. Given the impact of water supply disruptions, utilities may also be risk-averse and resistant to change.

In Singapore, PUB works with both the private sector and academia to commercialise innovations. It co-invests with organisations, and opens up some of its facilities for real-world tests, which are isolated from the rest of its network for safety.

Mr Ng said: "We hope that what we're doing in PUB will encourage and motivate others to embrace digitalisation too. The technologies are available, we have been successful in our projects, and digitalisation can solve many problems for all of us."

The following chapters will take a deep dive into PUB's digitalisation efforts, highlighting the deliberations behind its decisions, showcasing its projects, and underlining the infrastructural, cybersecurity and human elements that water utilities need to consider in charting their own paths forward.

Chapter 2

Why digitalisation?

As water utilities consider whether they should invest in digitalisation, they have to answer one decisive question: why do it?

For us in PUB, the road to digitalisation began with the confluence of two growing trends.

On one hand, Singapore's rising water demand, growing operational costs of supplying water, looming manpower shortages in the water sector, and new challenges such as climate change presented mounting problems in need of solutions.

On the other hand, advancements in digital and information communications technologies were transforming the global landscape, and could offer water utilities new methods of enhancing their productivity and efficiency in planning, operations and service delivery without greatly impacting costs.

In short, we saw that digitalisation held the answers to many of Singapore's incipient water issues.

By embracing digitalisation, we aim to become a smart utility that will continue to meet its mission of supplying good water, reclaiming used water and taming stormwater for people.

With that in mind, we came up with four major goals to guide our digitalisation efforts: to create value for the utility through new capabilities,

© IWA Publishing 2020. Digitalising Water – Sharing Singapore's Experience
Author: PUB, Singapore's National Water Agency
doi: 10.2166/9781789061871_0005

more efficient operations, a better work environment and improved customer service.

By setting out these four broad objectives, we have been able to focus our research and investments, plan our digitalisation journey in greater detail, and consider more deeply how to collect and use data in innovative ways. Data collected by one department within PUB would also be useful for others, while linking different data sets could lead to fresh insights to improve operations across the board.

CHARTING A JOURNEY

This higher-order thinking has also enabled us to set a general path for our transformation while leaving room to capitalise on creative projects that yield multiple benefits. As an example, we have placed sensors in Singapore's factories and sewers to detect illegal discharges.

The data is not only used for enforcement actions against errant firms, but also shared with water reclamation treatment plant operators, to give them more lead time to adjust the plant operations.

Between 2021 and 2023, we will also install 300,000 smart water meters in homes, commercial buildings and industrial sites across Singapore, as part of the first phase of the Smart Water Meter Programme. The smart water meters will provide near real-time water consumption information, replacing manual readings taken every two months. This endeavour, too, will reap many different dividends for both our customers and PUB.

With these smart water meters, our customers will be able to track their daily water usage via a customer portal, empowering them to be more water-conscious and use water more wisely. Our customers will also receive high usage or potential leak notifications and be able to fix leaks more promptly to reduce water loss and save money.

At the same time, our officers will be able to break the data down into daily or hourly readings, and into zones and sub-zones. By looking at the difference between water supplied to and used by consumers, our officers can find out where there is unaccounted-for water, enabling them to identify leaks and fix them earlier.

By leveraging digital technologies, we also plan to encourage behavioural change towards water conservation, optimise water demand management and achieve greater operational efficiencies. In addition, knowing demand patterns can help us to proactively plan our systems to better serve customer needs.

Deputy Chief Executive (Operations), Mr Harry Seah, noted previously: "All these digitalisation projects and others will make us more effective in our operations, save costs and result in fewer problems and complaints from the public."

ABSORBING LESSONS, MAKING ADJUSTMENTS

As we continue to implement digitalisation projects, PUB has also gleaned insights and many other valuable lessons. From our earliest projects, for example, we found that showing employees how digitalisation will improve their work-life balance is key to winning their support.

When organisations promote digitalisation, they tend to focus on benefits to the bottom-line, such as productivity gains. Employees, however, are more concerned with how digitalisation will affect their livelihoods and working environments.

They may resist digitalisation due to fears that they will lose their jobs to automation. Utilities need to address this, and show that digitalisation, far from threatening their employment, will actually boost their work-life balance.

By inviting employees to give feedback on proposed digital systems early and regularly, utilities can showcase how the systems will help them to do their jobs more effectively and potentially even shorten their work day.

At the same time, employees can also suggest crucial changes. While many people may expect digital systems to be perfect from the moment they are rolled out, that is rarely the case. Each system also has to be tailored to the specific needs of the employees who will use it.

For example, when a firm hired by PUB to design its Catchment and Waterways Operation System demonstrated the various dashboards to the different groups of operators who would use it, the operators requested further amendments that would enable them to carry out their job scopes better.

With every digitalisation project, we have also encouraged employees who benefit from it to share their experiences with their peers. Such word-of-mouth has done more to change minds and win hearts than the utility would have achieved with a management-led approach.

INTO THE FUTURE

As we look to the future, we see an increasing need for digitalisation. As cities aim to meet their residents' water needs while fulfilling other goals, such as limiting greenhouse gas emissions, they will have to become more nimble and tap into emerging technologies.

Through digitalisation and analysing data such as water consumption patterns, utilities can achieve real-time management of water, avoid over-producing water, make their treatment processes more efficient and save on resources, including electricity and chemicals.

We are already making advances on all of these fronts. Going forward, we will use digitalisation to step up the use of automation to replace laborious and repetitive tasks, so that it can free up manpower for higher value-added activities such as problem solving and developing new ideas. We are also leveraging artificial intelligence to support operations by reducing the time taken for decision-making and actions through analytics and automation. These strategies will alleviate a looming manpower shortage caused by Singapore's persistent low birth-rate.

Closed circuit television cameras, sensors and other surveillance technologies will act as eyes and ears to monitor water treatment plants, so that our officers need to do fewer physical checks, freeing up time for other tasks.

We will also automate repetitive and low-value work, such as scrutinising invoices for mistakes, so that employees can devote more time to higher-value responsibilities, and focus on improving their parts of the water system.

Digitalisation will also broaden our menu of options to tackle problems such as flash floods. In one instance in 2019, heavy rain had led to flash floods at Craig Road yet again.

While the conventional solution would have been to widen the drain at the road or build a culvert, analysis of data collected during the downpour and computer modelling pointed to a much less disruptive fix: constructing a weir.

Erecting the weir took just one day, and when equally intense rain poured in the area a few months later, there were no flash floods. With digitalisation, we had more informed options to choose from, and averted a tremendous amount of unnecessary work.

INTEGRATED WATER MANAGEMENT

Ultimately, our goal is to create a single efficient water system (Figure 2.1), where all of the processes that monitor water use, collect and treat used water, treat and produce drinking water and distribute it are optimised at the system level, rather than segregated based on function or department.

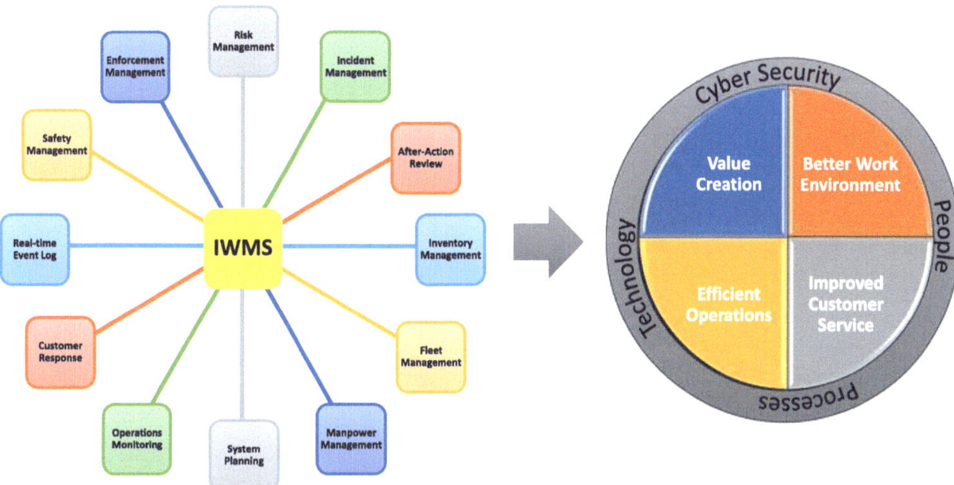

Figure 2.1 Core functions and outcomes of PUB's digitalisation efforts (IWMS: Intelligent Water Management System).

Human employees can harness digital systems to make this a reality. Both are essential for the journey and to reach the endpoint. While digital systems can solve many problems, they must be guided, operated and maintained by people with the right domain expertise. They are not a replacement for people.

Mr Seah emphasised: "Digitalisation is an enabler, but people have to make the decisions. Digital tools will give you more data to work with, more informed options and more confidence in optimising the system, but it comes down to people in the end. If we want water utilities to operate to the best of their ability, we need both people and digitalisation to work hand in hand."

Chapter 3

Digitalisation in PUB

When setting out on our journey towards digitalisation, our objectives were to create value through new capabilities, foster a better work environment, improve customer service, and generate more efficient operations. To this end, we have initiated projects across the length and breadth of PUB to achieve them.

In our transformation so far, we have created digital management systems, customised tools on the market for our specific needs, and launched research to engineer the next generation of digital products for water utilities. In many of these endeavours, we have benefited from the expertise of our partners in industry and academia.

Even as we continue to digitalise PUB's operations, our efforts have produced results and could be useful starting points for other water utilities to embark on their own digitalisation journeys. From crafting a roadmap to putting it into action, here's a look at how PUB has developed, tested and implemented digital innovations to achieve each of its four objectives.

CREATING VALUE

Every day, PUB manages a complex water loop that aims to collect every drop of water, reuse water endlessly and desalinate seawater to meet the needs of

© IWA Publishing 2020. Digitalising Water – Sharing Singapore's Experience
Author: PUB, Singapore's National Water Agency
doi: 10.2166/9781789061871_0011

homes, businesses and industries. While we have continually refined our strategies and technologies, digitalisation offers opportunities to bring our operations to the next level.

By tapping on digitalisation, we have incorporated new processes into established routines to make our work more efficient, found fresh ways to better monitor the health of our assets, and accomplished other breakthroughs. The following projects are some examples of how we have deployed digitalisation to create value in our operations.

Optimising the water system

Although Singapore has four national taps, namely water from local catchments, imported water, high-grade reclaimed water known as NEWater and desalinated water, it is up to our officers to determine how much of each tap to use to satisfy the nation's demands every single day.

These decisions are determined by factors such as the projected water demand, weather forecast, system constraints and the available capacity of each water treatment plant. Treating rainwater costs less than desalinating seawater, for example, so it is financially prudent to supply more water through the former than the latter if that is possible.

Throughout the day, PUB officers also monitor the water situation including weather changes, unexpected fluctuations in demand, and other possible disturbances such as plant failures. They have to make necessary adjustments to the plants' load dispatch to ensure that the demand is met and the water system is optimised.

To manage this process as smoothly as possible, our officers used to spend up to two hours at the start of each day planning how to fulfil Singapore's water demands while minimising costs and meeting operational constraints. They drew data on the operational situation from PUB's various departments and formulated a plan for the day based on standard operating procedures.

With digitalisation, this time-consuming process has been whittled down to a mere 15 minutes. An artificially intelligent program now analyses data collected by sensors and manual inputs by PUB officers to determine the best course of action for the day.

The system's ability to quickly process new data, cycle through "what-if" scenarios and continuously optimise the supply of water, has also given the officers the confidence to expand operational boundaries.

By lowering the service reservoirs' required start-of-day water level by 5%, for instance, we have cut our energy use for water treatment and pumping without affecting the supply of water to customers.

The digital system, jointly created by PUB and Singapore's Government Technology Agency (GovTech), generated significant savings for the utility in 2019, by optimising water transfers between reservoirs and the production of water to meet demand. It has also reduced the time that officers spend in planning operations by 64 man-days per year.

This implementation is just one example of how PUB has tapped into digitalisation to add value to its operations. PUB is also using digitalisation tools to create new value propositions, from a closer and more predictive look at the health of individual pump sets to smarter system-wide surveillance of the drainage grid.

Monitoring asset health

Pump sets are an indispensable part of every water utility. They keep water moving, inject chemicals into the water, and perform a range of other functions. PUB has more than 2,000 of them across Singapore, including transfer pumps at pumping stations, process pumps, and pre- and post-chemical dosing pumps.

Since 2018, we have been using low-cost vibration sensors by Proaspect Solutions at the Chestnut Avenue Water Works (CAWW) to test their effectiveness at monitoring and predicting the condition of the pump sets' (Figure 3.1). The sensors take readings on a regular basis and generate alerts via SMS to engineers when vibration threshold values are exceeded. These instances are further analysed to predict any impending component failure.

In addition, these sensors also monitor pump sets during start-up. Measuring and reviewing the pump sets' characteristics during the start-up phase may reveal any impending start-up problems or faults even earlier, to enable early action to prevent equipment failure.

To capitalise on this, we are embarking on a sensor analytics project with a local firm that will capture vital operating data from the pump sets, during both their start-up transient state and their steady state operations. It involves the creation of an artificially intelligent system that will analyse the collected data to detect signs of impending failure and identify the specific fault ahead of time.

Figure 3.1 Internet-of-Things (IoT) sensor installed on pump set for condition monitoring.

Such faults can include imbalances, misalignments, looseness, system resonance and rolling element bearing defects, as well as electrical induction motor faults like rotor defects and static eccentricities.

After an initial learning period, the sensors and analytical system will go into an evaluation mode where they will look for significant and sustained deviations in the pump sets' vibration behaviour from the established baseline.

PUB and Proaspect are also programming a low-cost system to analyse the electrical current signatures from the pump sets. By studying the captured phase voltage and current signals of the pump sets' low-tension motors below 600 volts, we will have another way to uncover imminent motor and pump mechanical faults.

Beyond preventing failures, a predictive maintenance sensor analytics system also improves maintenance regimes by allowing maintenance to be carried out on the pump sets based on the equipment's condition, rather than at fixed regular time intervals or based on their accumulated running or operating hours. This will optimise our maintenance costs and enhance operational reliability.

Apart from these projects, we have also been trialling smart sensors by ABB to monitor the condition of low-voltage motors in our plants. These sensors measure motor vibration, skin temperature, magnetic field and acoustic signals, and transmit the data for analysis.

Operators can also check the status of the motors via a mobile application, which colour-codes them according to their condition for easy review, with red indicating critical issues that need to be addressed immediately to prevent imminent failure, yellow signifying that the motor should be watched closely and serviced at the next possible opportunity, and green meaning the motor is fine.

Moreover, by combining data on individual motor energy consumption levels with plant operating information, it is possible to cut energy costs by adjusting appropriate motors. This project will also refine maintenance regimes, as with the pump set monitoring endeavours.

As we continue to harness digitalisation to better monitor the health of assets, factors such as the equipment's criticality, replacement lead time and cost will remain key to the decision-making process. Through these projects, we can pre-empt failures instead of reacting to them, minimising operational downtimes.

A smart drainage grid

As we move to reduce or even eliminate asset failures, we are also taking a wider look at managing our drainage system in a more pre-emptive manner. Since 2009, we have built up a hydrometric network consisting of more than 260 water level sensors and more than 80 flow sensors in waterways, and an array of closed-circuit television cameras along the waterways and across Singapore. We also have access to the National Environment Agency's rain gauges.

While PUB currently uses these instruments to issue flood warnings to the public in advance and alert staff to anticipate and manage floods better, more can be done with the data. By correlating these various strands of data, we can generate insights into how the various parts of the drainage system are performing, and thus improve our drainage planning and maintenance programmes on a strategic level.

For example, abnormally high water levels in drains and canals during smaller downpours, which were not observed for past downpour events of similar magnitude, may indicate choking or blockages downstream. By flagging these, we can investigate and resolve localised drainage problems via targeted maintenance, before they result in flooding during heavier storms.

By analysing water level trends at various waterways, and correlating that information to rainfall intensity, we can also identify drains that are undersized and may need more drainage capacity or alternative measures to prevent floods. Correspondingly, the drainage capacity of various sections of drains and canals would be known.

These insights are intended to be achieved as part of PUB's Smart Drainage Grid system (Figure 3.2), a data analytics system that consolidates all of the data generated by our sensor networks into a single database, that is updated in near real-time. To ensure that erroneous data is not used for analysis, the system automatically filters such data points, as well as flagging potentially erroneous ones which are then reviewed.

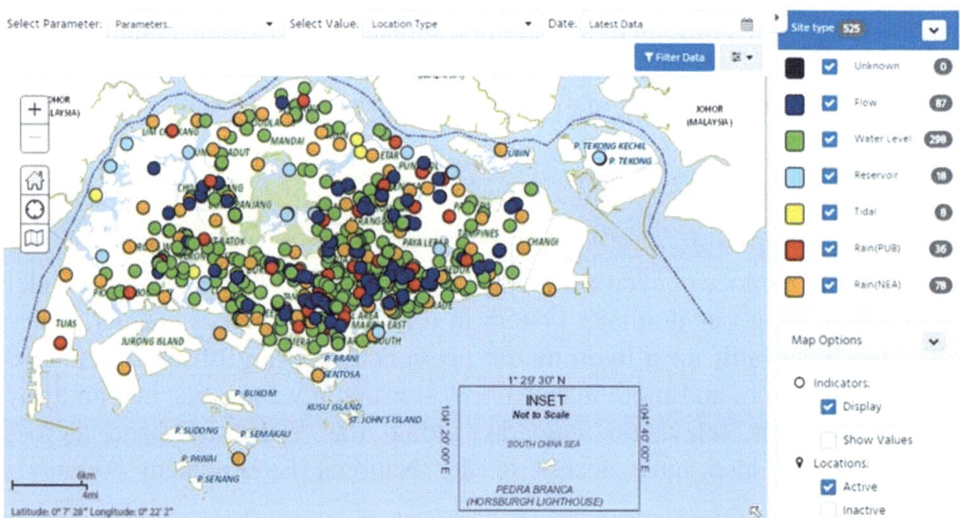

Figure 3.2 Map view of the Smart Drainage Grid.

The system was completed in June 2019 and is now operational, with ongoing enhancements from user feedback. PUB officers have been using the system for rain event-based analysis (Figure 3.3), to detect anomalous drainage performance for targeted maintenance to be carried out where necessary, and temporal performance analysis to generate insights for strategic drainage planning.

In addition to the examples as described above, PUB has also attempted to use the system to track the surface roughness of drains through the analysis of the

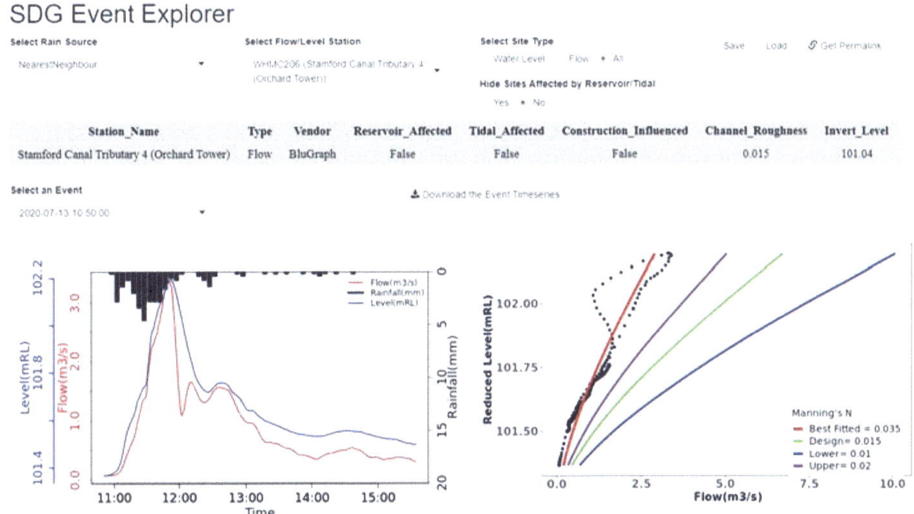

Figure 3.3 An example in the Event Explorer for the Smart Drainage Grid.

water flow-level response, as well as the rate of urbanisation in Singapore's catchments.

In the longer term, we can use the insights from these analyses to update our drainage planning parameters and norms in the Codes of Practice on Surface Water Drainage.

Detecting toxic pollution

While the drainage system harvests rainwater, a separate sewer network collects used water for treatment and reuse. PUB is digitalising the used water network too, to detect pollutants that are accidentally or illegally discharged into it. Such contaminants can harm the biological processes in a water reclamation plant and the production of NEWater.

We have imposed requirements on high-risk factories to install monitoring units for pH and Volatile Organic Compounds (VOCs) at their final inspection chambers to ensure that they comply with effluent discharge standards in Singapore. These monitoring units have deterred illegal discharges containing highly acidic or basic chemicals and prohibited VOCs.

To further protect the used water treatment system, we have worked with the National University of Singapore (NUS) to develop a Microbial Electrochemical Sensor (MES), and trial its use.

The sensor works on the principle of bio-electrochemistry, where high concentrations of toxic heavy metals and cyanide in used water cause the MES system's electrical voltage to drop, triggering an alert to designated officers.

Deploying these sensors will discourage factories from illegally dumping trade effluent containing cyanide and heavy metals, such as copper, nickel and zinc, into the sewerage system. They will also provide early warnings in the event of accidental or illegal discharges of these pollutants, so that we could act quickly and take necessary enforcement actions.

In 2018, NUS formed a spin-off company called EnvironSens to commercialise the technology. We are supporting EnvironSens through a larger-scale demonstration of 100 units of improved MES systems. Learnings from previous installations were incorporated into a re-engineered system, and maintenance regimes were optimised to drive down the cost of the technology with partial funding from the National Research Foundation. As of March 2020, 66 units have been installed at factory sites (Figure 3.4).

At the same time, EnvironSens is developing an Internet-of-Things platform to transmit and collate real-time data from the 100 MES units, and a web-based dashboard to present the data.

With the dashboard, our operators will have a map showing the status of the installed MES units, and will be able to access their individual

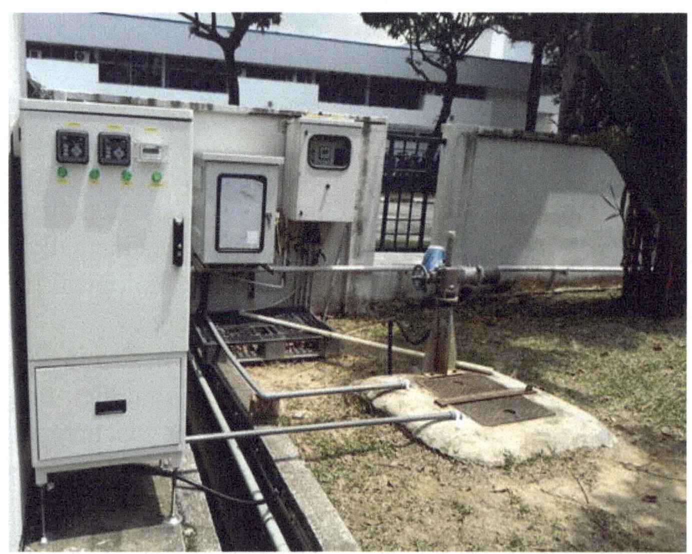

Figure 3.4 Image on setup of MES system at one of the industrial sites. (*Source*: EnvironSens Pte Ltd).

maintenance records, trigger records and other information. The MES network will also notify the operators through SMS and email when an alert is triggered in the MES units.

The novel MES system could even benefit water utilities beyond Singapore's shores. The real-time and online sensor is the first of its kind, and could find a huge market globally, including countries such as China, India, the United States, Vietnam, Australia and others.

With the four projects outlined above, as well as others that are ongoing or in the pipeline, we are transforming how we manage our water supply, drainage, used water and other systems. These digitalisation advances will prepare us to better meet new challenges, standing us in good stead for the future.

EFFICIENT OPERATIONS

Beyond creating new value propositions, PUB has relied on digitalisation in other ways to make its existing operations more efficient. By tapping on sensors, data analytics and predictive tools, we have revolutionised our management of the Marina Catchment drainage system and Marina Barrage, which are two key pieces of water infrastructure in Singapore.

We are also trialling a predictive control system to better manage a key part of our used water treatment process, and digital twins of treatment plants that offer engineers the ability to explore how various actions would impact plant operations before actually implementing them. These projects and their benefits are outlined below.

A computational breakthrough

The Marina Catchment is the largest in Singapore, with an area of 10,000 hectares that spans the most urbanised areas in the country. A combination of open-channel and closed drains make up the drainage system that flows into the Marina Reservoir, along with five major tributaries, namely the Singapore River, Rochor Canal, Stamford Canal, Geylang River and Kallang River, which includes the Whampoa River and Pelton Canal.

From 2015, PUB started collaborating with SUEZ Environment to build on their combined experience and construct a sophisticated system now known as the Catchment and Waterways Operations System (CWOS).

The CWOS fulfils three functions: optimising the operations of the Marina Barrage with an eye to maximising raw water storage while minimising flood risks in the city's low-lying areas; monitoring water levels in the upstream catchment; and monitoring and anticipating the quality of water in the catchment's reservoirs and waterways.

The CWOS assimilates more than 350,000 data points every day from various sources, such as rain gauges and water level sensors, as input for its operational dashboards and analytical and predictive models. To improve its accuracy, it has an algorithm that identifies and filters out anomalous data points that are likely to be errors.

The PUB-SUEZ team made separate dashboards for the CWOS's three main user groups, specifically staff in charge of drainage operations, water quality management and modelling, and the Marina Barrage's operations.

The drainage operations dashboard draws users' attention to areas with high water levels or rain gauge levels during downpours, so that they can best allocate manpower and response vehicles for such events.

Since the CWOS was deployed in 2017, it has provided actionable insights for over 170 storms, where rainwater accumulated to more than two-millimetres-high in the Marina Catchment.

The Marina Barrage operations dashboard uses current and forecasted conditions in the catchment to advise operators on the number of pumps or gates to be operated, taking into account constraints such as the operating range, delays or continuity. Its benefits have been validated in more than 150 situations.

The water quality management and modelling dashboard has three integrated water quality models that automatically render water quality forecasts. It also sends alerts when water quality sensors exceed pre-set normal ranges, so that water quality engineers can diagnose and solve problems quickly.

Apart from these dashboards, those using the CWOS can access a data explorer module to conduct more detailed analysis of time series data, generate reports and export data. The CWOS's ability to automatically produce and send reports, such as water quality daily reports and Marina Barrage post-event reports, has further reduced staff workload.

The automated reporting can even be customised for reservoirs and sent to specific recipients on a daily basis or after rain events. With this capability, reservoir management engineers no longer have to spend an hour each day logging into the system to extract water quality data, identify trends and prepare reports for the implementation of mitigative actions.

With the successful roll-out of the CWOS for the Marina Catchment, we are developing an integrated dashboard that will provide a holistic overview of its overall reservoir operations, as well as an individual dashboard for each reservoir's operations.

Among other improvements, the utility is also upgrading the water quality management and modelling dashboard so that the system can identify water quality issues, conduct source tracing island-wide and aid in inter-reservoir water transfer operations.

Advances in aeration control

With Singapore's water demand set to increase in the coming decades, finding ways to treat used water more cost-effectively is also key. We are piloting an advanced monitoring and predictive control system (Figure 3.5) in the Integrated Validation Plant, which was constructed to help the utility assess the feasibility of deploying new and innovative technologies in water reclamation plants and other PUB facilities.

The advanced monitoring and predictive control system enables us to better perform a crucial step in the used water treatment process: aerating the water so that ammonia can be removed from it.

This aeration control system has a self-learning algorithm that studies the ammonia load in the influent used water every 15 minutes. By collecting this

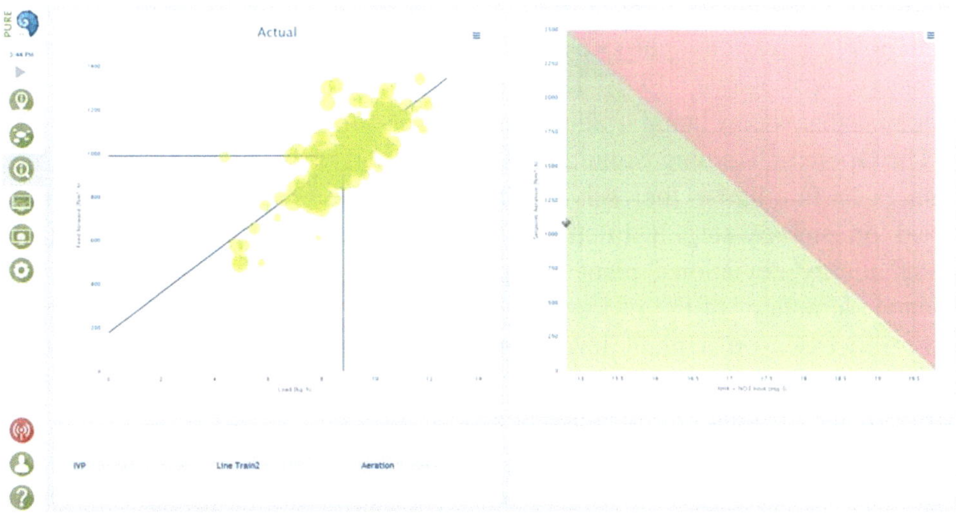

Figure 3.5 Aquasuite advanced monitoring and predictive control system.

data, it can establish and update the ammonia load pattern every day, and generate predictions for the load up to 48 hours in advance.

These predictions are fed into a second algorithm that determines the corresponding amount of aeration required for each train of the water reclamation plant.

To ensure greater accuracy, the prediction algorithm rejects sensor readings of the influent ammonia load that are probably errors, so that these do not derail its predictions. Through soft sensoring, it replaces such readings with historical operation data, and it also recalibrates itself by including data manually input by PUB staff about the effluent's ammonia load.

The pilot has shown that the system has a prediction accuracy of 88% and can reduce the amount of aeration needed by 15%, compared to the current approach where controllers take in readings from dissolved oxygen sensors and ammonia sensors to determine the amount of aeration needed.

By serving as an efficient autopilot solution for this part of the used water treatment process, the system can also help to address the potential shrinking of the pool of available skilled operators in the future.

Digital twins

While predictive systems offer a glimpse of optimised systems in the future, simulations are a safe space for operators to experiment with "what-if" scenarios for various purposes. With Japanese engineering and software company Yokogawa, PUB designed an integrated system with three functions that is now being trialled at the Lower Seletar Water Works (LSWW).

The Operator Training Simulator (OTS) supports self-learning which allows trainees to familiarise themselves with the plant's processes and operations based on step-by-step instructions for various malfunction scenarios (e.g. water quality deviations, plant equipment failure). The core of OTS is the dynamic simulator of the process which features rigorous models, can handle large processes on a plant level, and can perform high-speed simulations. OTS also allows for instructors to lead, monitor, and evaluate the trainee's performance through an instructor platform.

Two other functions, called the Logic Check-out Platform (LCP) and Mirror Plant System (MPS), enable maintenance and operations engineers to study, modify, test and validate process and treatment changes, before applying them in the plant.

Currently, our engineers manually calculate the effect of proposed process configuration changes, and may unintentionally miss out interactions between tightly-coupled subsystems due to the complexity of the water treatment processes.

The LCP delivers wider, time-varying and accurate simulations of the results of process changes. With it, our engineers can better evaluate modifications that are proposed for maintenance or operational needs, and react quickly to any potential consequences of cybersecurity attacks.

As a water treatment plant's control system and configuration will need to be updated continuously throughout its lifecycle to cater for new equipment, modified control algorithms, better user graphics, enhanced safety logics and other developments, the LCP will aid in ensuring that these changes are carried out without disrupting the plant's operations.

The MPS, on the other hand, demonstrates the outcomes of water treatment input changes. Water treatment depends on several factors, including the raw water's quality and the chemicals used to treat it. Currently, programmable logic controllers adjust the treatment processes according to the quality of the incoming raw water, but this is a reactive approach.

Plant operators can be more proactive with the online MPS, which is a digital twin running in parallel with the actual plant. It collects near real-time data from the plant's local replicated historian server, and simulates treatment processes.

As the MPS has detailed information about the plant and its operations, including the size of equipment and flow rates, operators can use it to accurately see how various actions would affect the plant's water treatment performance before they apply the best operation method.

As an example, part of the water treatment process requires dosing raw water with a chemical. Plant operators need to decide not only how much of the chemical to use, but also when to use it. By running "what-if" scenarios on the MPS, operators are able to determine the optimal amount and time to carry out dosing.

By providing digital replicas of water treatment processes and scenarios, the integrated system can strengthen our operators' maintenance and operational skills. As with the other featured projects, such digitalisation will help to make our operations safer and more efficient.

BETTER WORK ENVIRONMENT

As sensors, robots, drones and other technologies become more sophisticated, they can help water utility operators by carrying out repetitive and labour-intensive tasks, or work in environments too hazardous for human employees.

Over the years, we have developed machines and systems to lighten our workload, arm our staff with better tools, keep them safe and upgrade their skills. This investment has created both a better work environment and happier workforce, and will also prepare the utility for a future where Singapore's low birth-rate and growing economy may lead to manpower shortages.

The following projects are a snapshot of our efforts to improve the workplace. From a smart Unmanned Aerial Vehicle (UAV) for tunnel inspections, to a Radio-Frequency Identification (RFID) system to track spare parts, these demonstrate the breadth of possibilities for water utilities.

Unmanned inspections

In 2008, PUB completed the first phase of its Deep Tunnel Sewerage System (DTSS), which is an underground superhighway for used water management. The construction for DTSS phase 2 is underway, and when completed in 2025, DTSS phase 1 and 2 will comprise a network of trunk sewers leading to two major tunnels that criss-cross Singapore to transport used water to three centralised water reclamation plants.

A 48-kilometre-long deep sewer tunnel was constructed during the first phase to convey used water by gravity from Kranji in the north-western part of Singapore to Changi in the east. The Changi Water Reclamation Plant (CWRP) sited at the end of the tunnel treats about 900,000 cubic metres of used water daily.

To ensure that the DTSS continues to convey used water to the CWRP for treatment, it is necessary to carry out regular inspections on its corrosion protection lining, comprising a High-Density Polyethylene (HDPE) lining and a cast-in-place concrete lining.

As the tunnel is located deep underground and carries used water, hazardous vapour is present within the tunnel and it is dangerous for humans to enter for inspection. It is also impossible to shut off or isolate the tunnel, even for a short period of time.

To overcome this problem, we worked with UAV developer AeroLion Technologies to design and produce a UAV that can fly in the tunnel over the water and conduct the inspections in their stead.

While many UAVs exist in the market, we needed one with customised capabilities, as the tunnel is long, with space constraints and completely dark, with its entry and exit shafts situated in remote locations. In addition, Global Positioning System signals cannot be received in the tunnel.

In the first stage of the UAV project, the team custom-made a UAV platform with on-board sensors, including navigation cameras, omnidirectional inspection camera, range finders, accelerometer, gyroscope, barometer, magnetometer and methane gas detector (Figure 3.6). It also has illumination LED lights to provide lighting for its cameras in the tunnel's pitch-black environment.

Beyond the hardware, the team devised real-time three-dimensional navigation algorithms that would work based on the navigation cameras' sensing, and programmed them into the UAV's on-board miniature computer.

A launch and retrieval mechanism was also invented so that the UAV can be safely launched and retrieved from the bottom of a vertical shaft. In an inspired touch, the mechanism also acts as a communication relay between the UAV and an external Ground Control Station to solve the non-line-of-sight issue.

The UAV has been flown successfully into the main DTSS tunnel via its shafts. The team is currently improving the UAV's location tagging accuracy

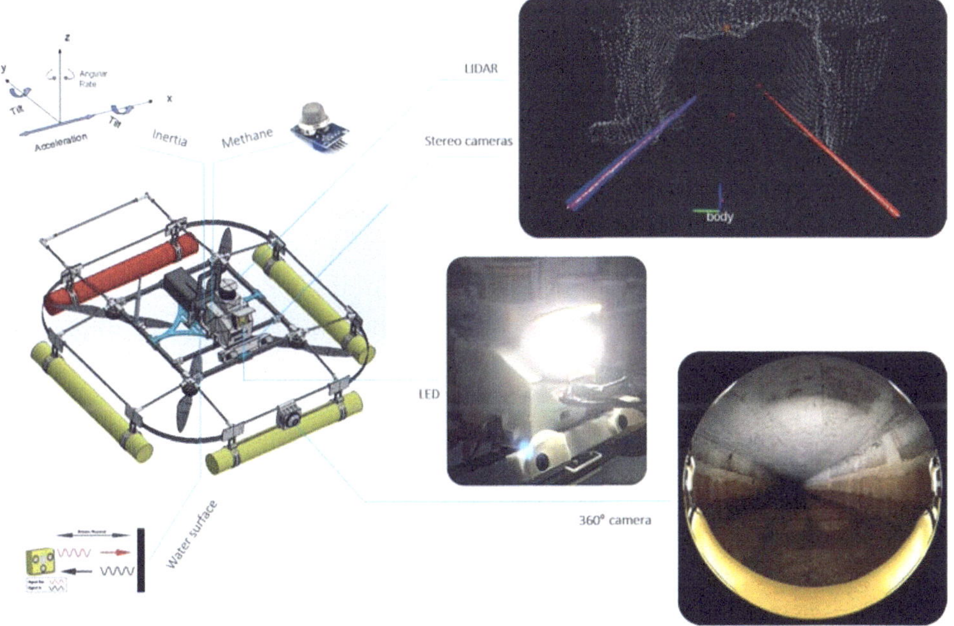

Figure 3.6 Design of the UAV for DTSS Inspection. (*Source*: AeroLion Technologies Pte Ltd).

and expanding the variety and quality of its data collection. Part of this will involve adding more gas sensors and a Light Detection And Ranging (LIDAR) sensor.

In addition, plans are underway for more advanced image and data analytics to automatically generate and visualise the UAV's inspection results, with adjustments to make it stronger, more efficient, and more convenient to use. When ready, the state-of-the-art UAV will be a key component of PUB's maintenance toolbox, enabling the utility to carry out man-less inspection in the tunnel.

Automating tests

Apart from taking the place of people in potentially hazardous environments, robotics and other digitalisation tools can also shorten the time that water utility operators spend on repetitive and manual work.

To meet Singapore's growing water demand, we must plan ahead and expand our water supply infrastructure and networks. This will in turn require the testing of more water samples. To cope with the higher workload, we have introduced a manufacturing execution system to integrate the various testing processes and automate them, from data collection and sample preparation to the actual testing and reporting of the results (Figure 3.7).

Previously, PUB staff had to manually prepare the samples and test them on five different instruments for six parameters. Now, the manufacturing execution system receives instructions from the existing Laboratory Information Management System (LIMS) on which tests to carry out for each sample, executes the tests, and sends the results back to the LIMS.

This change has given staff more time to focus on other work. The manufacturing execution system has also eliminated human errors and enabled tests to be conducted around the clock, greatly improving productivity. The test results can also be more easily traced through the system's software.

Furthermore, the manufacturing execution system was designed and constructed to allow for potential modifications and expansions. For instance, an additional online analysis module can be supplemented to double the rate of analysis. More parameters can also be added to the system to maximise its testing capability. By instituting this system, we will be able to handle more tests cost-effectively, and ensure that the results are reliable, accurate and traceable, in the long run.

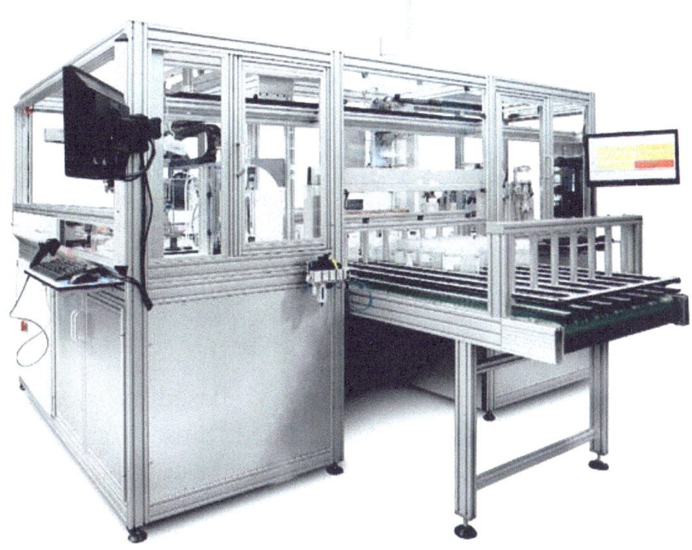

Figure 3.7 The TPUB Water Analysis System. (*Source*: Labman Automation Ltd).

Tracking spare parts

A similar effort to automate the counting and tracking of spare parts has also been very successful. Keeping a well-stocked supply of spare parts for different types of maintenance work is crucial for PUB's mission, but took up an inordinate amount of time in the past.

Officers had to manually record the details of each spare part and its movement – from arrival to disposal. They also needed to set aside a large amount of time for the monthly stock-taking exercises. The manual management of the spares was cumbersome, and occupied hours that could be better spent on other tasks.

Digitalisation technologies such as RFID systems are capable of carrying out the tasks in a more efficient manner (Figure 3.8). PUB trialled an RFID system at Choa Chu Kang Waterworks (CCKWW), which has more than 30,000 spare parts in its store.

The parts were tagged at either item or batch level. Large parts, such as motors and pumps, received individual RFID labels. Batch items, such as nuts and bolts, are kept in trays, and an RFID label was affixed to each tray. RFID gantries, and wall and mat antennas, were also installed as part of the system.

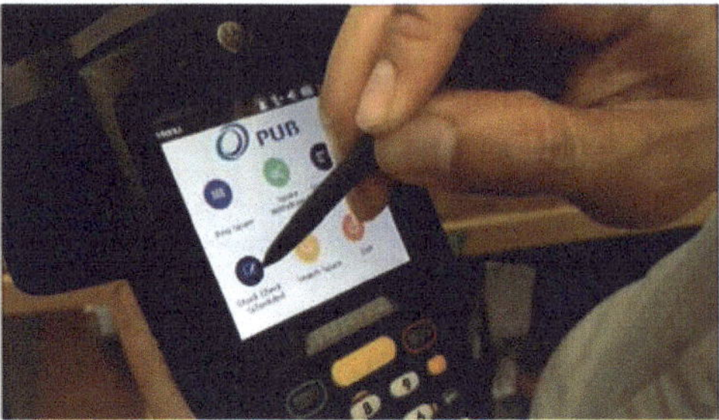

Figure 3.8 Inventory check using RFID system (2 images).

The trial also involved customising a web-based program with online forms and specific modules, which included those for registration and withdrawal of parts. Supervisors could approve these requests online, get an overall view of the store's supplies at any time, and be notified when stocks fell below predetermined levels. These functions led to better inventory management.

The RFID system also allowed for a more efficient stock-take process. With the system in place, our officers slashed the time needed by up to 92%. This included the time spent on transferring data, which not only made the process faster but prevented errors as well.

The RFID system also had other benefits aside from streamlining the management of the spare parts and making the various processes much more efficient. With the gantries in place, it could detect unauthorised withdrawals of items from the store and alert officers.

With the CCKWW pilot validating the usefulness of the RFID system, we have integrated it into our wider Asset Management System (AMS).

Health and safety first

As the technologies behind wearable smart devices mature, they can also revolutionise other aspects of water utilities' operations, and make work easier and safer for our staff. When working on-site at water treatment plants and other facilities, PUB employees use a buddy system to ensure their safety. This arrangement, however, is not an effective use of their time.

The utility is trialling wearable safety trackers that would replace the buddy system without compromising on safety (Figure 3.9). These devices help track our employees' locations through GPS and geo-fencing, both indoors and outdoors in real time, and have distress buttons that can be

Figure 3.9 Wearable safety trackers.

pressed in case of emergencies to alert the plant's control centre to send immediate help (Figure 3.10).

Taking advantage of this location-tracking capability, virtual geo-fencing can also be applied to different areas in the treatment plants to spot unauthorised loitering or wandering around in buildings and process areas. Each tracker

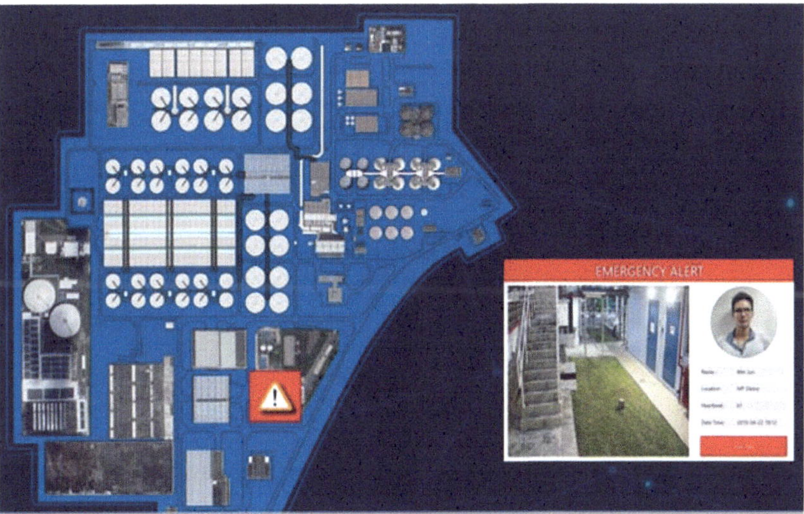

Figure 3.10 Geo-fencing intruder alert system.

can also double up as a token to grant or restrict the employee's access to fenced areas or secured rooms such as electrical switch rooms and process unit tanks.

The pilot project at the Ulu Pandan Water Reclamation Plant showcased the trackers' effectiveness but also highlighted potential obstacles to their deployment, including the relative infancy of low-cost wireless solutions for indoor uses. PUB is now in the midst of a larger trial involving more than 150 trackers for its staff and contractors in the Ulu Pandan plant and Chestnut Avenue Waterworks.

Through this trial, we will evaluate various modes of low-power wireless communication, and further study the process management and logistical aspects of implementing the trackers, with a view to using them in all plants.

Augmented reality

PUB has also created an augmented reality application that runs on Microsoft's HoloLens to digitalise maintenance processes.

When our officers carry out troubleshooting, routine servicing or corrective maintenance work, they usually bring along some form of documentation, such as printed manuals or photographs on their handphone, to refer to on-site. Finding the relevant information can be time-consuming, tedious and inconvenient, especially when they must go through the materials page by page while completing the task at hand.

With the augmented reality headset, they can easily access and view equipment's technical documents and specifications on-site. They can share their field of view with these remote colleagues and experts, who can in turn project technical documents and images, and even draw, on the operators' field of view.

The application also has animated demonstrations of maintenance procedures, complete with three-dimensional models, to guide new or inexperienced staff on how to execute them safely and properly (Figure 3.11). This function will be invaluable in training staff and refreshing their skills.

While documents and drawings for older equipment and plants had to be digitalised from their original paper format, this is unlikely to be necessary for newer generations of equipment and plants. Additionally, as augmented reality-enabled headsets and glasses gain more powerful processing

Figure 3.11 Use of AR to support on-site maintenance work.

capabilities from hardware improvements, they will be able to support more sophisticated visualisation applications.

The diversity of these featured projects shows that water utilities can harness digitalisation in a variety of ways to enhance employees' ability to perform their duties at work. Far from being a threat to their livelihoods, digitalisation can help utilities to look after employees' well-being, and aid them in discharging their duties more quickly and effectively, empowering them to achieve a better work-life balance.

IMPROVED CUSTOMER SERVICE

Ultimately, a water utility's goal is to serve the people. To better carry out its mission, PUB has also initiated digitalisation projects to improve our customer service, solicit more feedback and provide our customers with more real-time data to help them to curb their water consumption.

Some of these innovations, which include the widespread installation of smart water meters and the deployment of robots to patrol key installations and reservoirs, are showcased below.

Smart meters for a smarter system

Between 2021 and 2023, as part of the first phase of the Smart Water Meter Programme, we will install 300,000 smart water meters in new and existing homes as well as commercial buildings and industrial estates. These smart meters will transmit near real-time water consumption data, as compared to the manual meters that are currently manually read once every two months.

With this near real-time data, we will be able to identify and fix leaks more quickly, optimise our water production and network management by matching supply to demand, among other advantages.

Customers will benefit too. They will be able to track their daily water usage via a customer portal, empowering them to be more water-efficient. Customers will also receive high usage and potential leak notifications, enabling prompt action to fix any leaks.

PUB can also illustrate to customers how their water usage measures against others and similar household units. Both self-comparisons and peer-to-peer comparisons have been found in studies to lead to significant decreases in water usage. PUB will design the means, such as customer portals, to push such data and enable comparisons for customers.

Apart from furnishing comparisons, we are also looking at gamification as a potential way to encourage water conservation. By turning water conservation into a game, complete with virtual points for achieving specific goals, more people may be motivated to use water more efficiently.

Such gamification techniques are also more likely to draw the attention of teenagers and children, compared to more traditional water conservation campaigns. People who cultivate water-saving habits from a young age are more likely to practise them throughout their lives.

PUB has already validated many of these measures in two smart water meter trials conducted in 2016 and 2018, respectively. The 800 participating households reported an average of 5% water savings due to earlier leak detection and adopting good water-saving habits.

Ensuring security with robots

Beyond water operations, we are also responsible for maintaining the security of our public reservoirs and key installations.

PUB's reservoirs are frequented by Singaporeans and visitors for their scenery and a host of activities, such as water sports and fishing. To safeguard reservoir water quality and ensure the safety of visitors, our officers regularly conduct patrolling and surveillance duties at the reservoirs.

Over at our key installations, PUB outsources its security needs to a security firm, which provides manpower and surveillance technologies, such as closed circuit television systems, that are monitored at the Integrated Operations Centre.

This arrangement has become increasingly difficult to maintain because of the security sector's growing challenges in hiring and retaining people. Furthermore, the costs of training the employed security guards to the required standard have increased over time. This means that the existing security model which is manpower-intensive may become unsustainable in the long run.

Digitalisation, specifically in the form of intelligent security robots for patrolling and surveillance, may be a long-term answer to this problem. We are working and partnering with security vendors to develop robots to complement and augment current patrol and surveillance operations. One such robot which is currently under development is the O-R3, an autonomous mobile robot equipped with video and audio capabilities.

Unlike human security guards, the O-R3 is able to patrol the area continuously without becoming physically tired or mentally fatigued. It can move autonomously and avoid static and moving obstacles, and has various functions for security work, including a 360-degree camera system for live video streaming and real-time data collection, and video analytics to detect anomalies.

PUB has successfully tested the O-R3 in two different environments: Bedok Reservoir, which is open to the public, and Lower Seletar Waterworks (LSWW), which is a restricted area. Bedok Reservoir has many visitors, including children, joggers and cyclists, whereas the human traffic at LSWW is minimal and consists mostly of PUB's staff and contractors. The two separate trials tested the O-R3's capabilities in two very different environments, each having its own challenges and issues.

At Bedok Reservoir, the O-R3 has been tested to augment PUB's reservoir surveillance efforts to deter illegal reservoir activities when members of the public go near the reservoir edge (Figure 3.12). The public embraced the O-R3 during its trial patrols at Bedok Reservoir, with many people taking pictures and videos of it, and sometimes even with it.

Due to the recent outbreak of COVID-19, safe distancing measures have been enforced to reduce movements and interactions in public and private places, to minimise spread of the virus. In response to public feedback on non-compliance with safe distancing measures at the reservoir parks at Bedok and Pandan Reservoirs, we also saw an opportunity to deploy the O-R3 to

Figure 3.12 O-R3 on Duty at Bedok Reservoir.

operate as a Safe Distancing Ambassador. The robot patrols around the reservoir and broadcasts messages in four languages, reminding visitors to refrain from gathering or loitering at the reservoir parks and to adhere to safe distancing requirements.

The eventual goal of deploying these surveillance technologies, namely robots, off-site monitoring, may seem like a distant reality today, but it may be the reality in future.

Automated alerts for flood risks

As part of its comprehensive drainage management and control system, we have installed over 200 sensors in key canals and drains to track their water levels. While such digitalisation has improved our operations, we have also used the sensors to offer the public a flood risk alert service, via SMS or PUB's MyWaters app.

People can subscribe to get alerts from up to three of the sensors. They will be notified when the water level rises above 50%, 75%, 90% and 100% of the selected waterways' depth, and when it drops below the respective alert levels.

With this automated system, subscribers will be forewarned, and can take precautions against, potential flash floods at the earliest possible time. Such alerts are a prime example of how water utilities can use digitalisation to improve customer service.

As the sheer variety of projects featured in this chapter demonstrates, water utilities have many options when it comes to digitalisation, and they stand to reap gains in customer service, productivity, and many other areas. Incorporating digitalisation will make a utility's operations smarter, and is the smart choice.

Chapter 4

The digitalisation framework

With each of its digitalisation projects, PUB has gained valuable experience that can be applied to our future efforts and help other utilities in their digital development. Our journey to date has highlighted the importance of a few key considerations and principles, in particular, that should underpin the digitalisation journey.

To digitalise their operations, utilities must have the technologies in place to collect data, infrastructure to transmit and share it, and analytics to make sense of it and turn it into actionable insights (Figure 4.1).

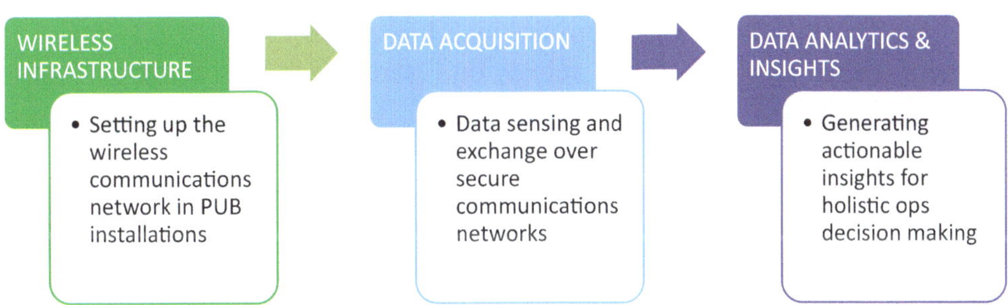

Figure 4.1 The digitalisation process.

© IWA Publishing 2020. Digitalising Water – Sharing Singapore's Experience
Author: PUB, Singapore's National Water Agency
doi: 10.2166/9781789061871_0037

As every digitalisation project is different, choosing the right combination of tools each time is crucial. For long-term use, the hardware and software should also be designed with continuity in mind, so that utilities are not locked into any particular protocol or vendor.

PUB has made these and other fundamentals part of its digitalisation framework for crafting and executing projects. This framework, detailed in the following sections, could be useful for other water utilities.

BUILDING THE RIGHT WIRELESS INFRASTRUCTURE

An appropriate, robust and secure wireless communications network is one of the most important components in digitalisation. Without a reliable and secure way to transmit data, all of the information collected by sensors, drones, robots and other technologies has no way of being analysed and operationalised.

There are many communications networks in the market today that cater for different uses and applications, such as 4G, Wi-Fi, LoRa, SigFox, etc. When deciding which network to use, water utilities should consider the required network range and transmission rate, size of data to be transmitted, the network's power consumption and latency, and other factors.

Security should be a pivotal concern. our systems are governed by data classification frameworks and stringent government cybersecurity policies, especially for systems that are within Critical Information Infrastructure (CII) installations. We therefore only deploy wireless infrastructure that meets the mandated security standards.

In our digitalisation projects to-date, we have used different networks for different purposes. We have relied on NarrowBand Internet-of-Things (NB-IoT) networks for outdoor positioning, and Bluetooth beacons and Bluetooth Low Energy (BLE) gateways for indoor positioning.

At the Integrated Validation Plant (IVP) located in Ulu Pandan Water Reclamation Plant, we built a BLE network for asset conditioning monitoring because of its low cost, ease of deployment and ability to facilitate real-time monitoring of assets for predictive and preventive maintenance. We also constructed a LoRa network to support the tracking of staff and contractor locations via smart watches as the network offers coarse positioning based on signal strength at relatively lower cost.

PUB also installed a Wireless Process and Environmental Sensing network to monitor pipe corrosion, hydrogen sulphide gas, and the level and flow of streams in sewerage networks because it provides reliable communication with radio frequency mesh routing capability, and overcomes interference issues in plant environments with dense infrastructure.

In some cases, we have turned to hybrid solutions. Networks may not be able to achieve full coverage within treatment plants, especially in some indoor areas and basement levels. These may become dead zones, preventing devices within them from transmitting data.

As an example of how PUB overcame this problem, we used NB-IoT for outdoor positioning and a BLE-based solution for indoor positioning, both integrated into the same wearable tag used for personnel tracking, in addition to increasing the number of repeaters and gateways.

We are also looking ahead to anticipate our wireless communications needs and opportunities. From 2020 onwards, we will deploy Wi-Fi networks in our water treatment plants to ready them for potential IoT applications such as smart wearable devices, equipment sensors for asset management and robotics surveillance.

We are also identifying bandwidth-intensive applications, such as virtual and three-dimensional augmented reality applications, drone surveillance and control operations at reservoirs, and real-time video streaming analytics, that could benefit from the 5G network that is expected to be rolled out across Singapore starting in 2020.

HARVESTING DATA

With technological advancements in recent years, water utilities are also now spoilt for choice when it comes to data collection devices. PUB has tried, tested and used many of them, and is trialling many more, including the vibration sensors to monitor and predict the condition of pump sets covered in the previous chapter.

PUB has established a smart water grid system called WaterWiSe. The development and implementation of the near real-time system was initiated in 2013. Today, there are 346 sensor stations island-wide to track in real-time the pressure and quality of water, including pH value, conductivity and turbidity, in the potable water network. By the end of 2020, we will install another 130 sensors and expand the coverage to our NEWater and industrial water networks.

By keeping an eye on the pressure in the network in real-time, the sensors have eliminated the labour-intensive work of installing temporary pressure loggers, pre-empted problems such as potential leaks, and detected pipe bursts, which generate pressure transients.

By 2025, we will integrate the sensor data with data from the upcoming smart water meters network, to create virtual district meter areas for better leak detection. This will also aid in the dynamic forecasting of network events.

We are also investigating other types of data collection devices. Currently, we send staff out on small vessels to patrol reservoirs. This is labour-intensive and provides only limited spatial coverage. Furthermore, water quality monitoring at reservoirs is done via stationary profilers and the manual collection of water samples for laboratory tests, but both methods offer only information about the water at the sampling locations.

On the other hand, drones can be used to carry out aerial surveillance operations. They can monitor large water bodies quickly and analyse images to flag anomalies such as algal blooms. They can also fly over pipelines to spot leaks and inspect confined spaces such as sewer tunnels. Drones equipped with cameras to capture illicit activities, hyper-spectral cameras to monitor water quality, and thermal cameras to discern water leaks could keep watch over much larger areas with far more efficiency.

Beyond sensors and drones, robots can collect data too. Apart from the patrol and surveillance robot featured in the previous chapter, we are also utilising remotely operated vehicles for underwater inspections and robots for the inspection and repair of big sewers.

BREAKING DOWN SILOS

Collecting data, however, is just the first step. As in other complex organisations, departments in water utilities may harvest data for their own use and fail to share it with other departments. Breaking down such information silos is crucial as data gathered by one department may be useful to others, and consolidating all of the data could lead to new correlations and practicable ideas.

PUB's ongoing project to upgrade our Intelligent Water Management System (IWMS) is a prime example of this. Currently, the system integrates some data from key operations systems and is mainly used for rudimentary operations monitoring and reporting.

The improved IWMS will integrate both operations and non-operations systems to analyse data from across the water loop to enhance our service and engineering operations. This will enable the IWMS to serve as the underpinning system for operations at our Joint Operations Centre (JOC), which is the nerve centre for all real-time and on-demand critical operational decisions.

The IWMS will also have new functions that demonstrate the possibilities that arise from aggregating data. For example, the IWMS's new incident reporting function will replace the existing email reports submitted by our officers. It will allow officers to report incidents and subsequent updates through their mobile devices. Our officers will be able to pin locations, attach documents and capture images while managing the situation at the incident site. We will also have automatic approval workflows to streamline incident report submissions.

With this new function, our officers will be able to track incidents efficiently from initial reporting to escalation to closure. Information captured in the module will also be automatically available for analysis and review in the After Action Review Portal.

During each incident, all of the information necessary to manage it will be consolidated and presented on a single Operations Dashboard for the convenience of the officers-in-charge, the JOC and our senior management. Our officers managing the incident site will be able to share real-time status updates to the dashboard via mobile devices.

If there is a potable water supply disruption, for instance, information such as the relevant pipeline pressure sensor data, indicative leak location, number of affected customers, PUB resources at the incident site, and real-time updates from the response teams will be displayed on the dashboard.

In the meantime, a Customer Response function will automatically assign resources to incidents based on the nature of the incident, the officers' expertise and location. The JOC will no longer have to manually deploy resources, resulting in more efficient assignments.

PUB officers responding to customers will be better equipped with information such as the customers' profiles, support resources' location and time of arrival, similar cases in the vicinity, checklist for investigation and other data through their mobile devices. This will empower them to better serve the affected customers.

After incidents occur, the After Action Review Portal will enable the review audit of operations, as well as capture findings, track open items and remind departments to follow-up. It will also facilitate further analysis, such as insights into the frequency of specific types of incidents.

SUPPLEMENTING DATA

As PUB works to centralise its data and maximise its use, we are also supplementing the data with soft sensors to mitigate instrument errors.

For example, in used water treatment, process automation is highly dependent on instrument readings. Errors in the instrument readings could lead to problems in the treatment process and poor effluent quality.

While maintaining the instruments would help to prevent errors, keeping them pristine is easier said than done. Dirt and micro-organisms found in used water can cause instruments to clog and foul. Electronics are also corroded by the hydrogen sulphide gas generated from used water.

Soft sensors would help to mitigate faulty readings from instruments. These are not physical sensors, but inferential software models that use historical data, easily measured variables, or a combination of the two to estimate process variables.

Take water flow rates as an example. A soft sensor can derive the flow rates by looking at the number of running pumps, the pumps' speed and their energy consumption. It can also produce dissolved oxygen estimates by analysing data related to incoming mixed liquor suspended solids, the influent flow rate and distribution, influent ammonia level, valve positions and air pressure.

Having soft sensors would help water utilities in many ways. If an instrument fails, its readings can be replaced by soft sensor estimates. Growing disparities between the estimates and instrument readings would also alert operators to issues, either in the instrument or in the corresponding treatment process, that need to be addressed.

In a trial at the Changi Water Reclamation Plant (CWRP), we tested the accuracy of soft sensors for dissolved oxygen against their physical counterparts. Before maintenance was carried out on the physical sensors, the soft sensor estimates were more accurate. After maintenance, the readings were similar.

The only limitation of soft sensors is that their accuracy would drop after a certain amount of time, as they ultimately rely on physical sensors, which, as explained, are difficult to maintain in mint condition. Even if soft sensors make use of historical data, if the physical instruments supply erroneous readings over a prolonged period, their accuracy will inevitably diminish.

For this reason, soft sensors are best used for detecting faults, and as a source of replacement data for short periods of time. Based on these merits, we will adopt soft sensor systems in the CWRP and IVP in late 2020. If these are successful, they will eventually be used in all water reclamation plants in Singapore.

TRANSLATING DATA

After data is collected and verified, it needs to be analysed and turned into useful insights that can be acted upon. This is why water utilities must also invest in enablers for process and system optimisation, such as video analytics, machine learning and digital twin technologies.

To digitalise security at PUB's waterworks and water reclamation plants, we are commissioning a system that will apply video analytics to video camera feeds. The intelligent system will have facial recognition capabilities that are powerful enough to identify more than five people per frame, and do so in under 10 seconds.

With these capabilities, the system will be able to flag unauthorised access to areas or buildings, or detect loitering within restricted areas.

Upon detection of movement within a restricted zone, the system can instruct cameras to stay focused on the source of the movement, whether it is a person or vehicle, and track it until it is out of sight. These camera live feeds will be grouped and displayed on the camera monitoring workstation for security attention.

Our staff can also use a search function to isolate all recorded footage that contains a specified attribute, such as moving vehicles or a particular face. They can then play the results according to the time sequence and track the movement of a person or a vehicle through different camera views, or display the results in thumbnail format.

By spotting personnel who are not wearing safety helmets or vests – the colour of which can be regulated by PUB to facilitate such detection – the system can also be used to enforce safety rules in the installations.

In the future, we may also expand the system's functionality so that it can uncover defects in pipes and solar panels, observe the progress of construction, pinpoint leaks via thermal imaging, and handle other responsibilities.

While this video analytics-powered security system is under development, we are already using image analytics technologies in other operations.

Our Silt Imaging Detection System utilises image analytics to turn CCTV cameras into sensors. The CCTV cameras are located at construction site discharge points to public drains, and the analytics software alerts staff to changes in the colour of the site discharge.

This digitalisation upgrade saves our contractors more than 100,000 man-hours every year and over $500,000 annually for the utility. It has also increased the capturing of non-compliance occurrences.

PUB's Fish Activity Monitoring System (FAMS) depends on image analytics technology too. We use fish as a means of monitoring water quality at waterworks and service reservoirs to ensure a safe supply of drinking water. FAMS analyses the fish motions and alerts operators when it finds signs of distress or abnormalities that could indicate changes in the water quality.

THE POWER OF PREDICTIONS

With data analytics and machine learning, water utilities can also produce predictions based on available data and take a more proactive stance in operations. While most utilities now fix problems after they are alerted, with more accurate predictions, utilities could be forewarned of potential issues and even prevent them from happening. In this spectrum, PUB has partnered with different solution providers to deploy data analytics and machine learning tools to aid the predictions in its vision of smart plants and smart networks.

In used water treatment, we have worked with Royal HaskoningDHV and ST Engineering Marine to trial the Aquasuite PURE software to improve effluent quality while reducing energy consumption, and to integrate ST Engineering Marine's Sensemaking algorithm for predictive maintenance into Aquasuite PURE.

Aquasuite PURE predicts used water flows, monitors performance and controls key treatment processes by using artificial intelligence and machine learning that predicts influent flows and continuously optimises performance.

The Sensemaking software employs a proprietary data analytics algorithm to perform condition-based monitoring on platform machineries, and it also has a decision support engine to provide recommended actions to operators. It uses various analogue parameters to analyse and learn from past failures to predict the time to the next failure, calculates the effective operating hours of the equipment, and gives a recommended time to the next maintenance, which may differ from the scheduled maintenance.

Leveraging on the capabilities of data analytics and machine learning for water treatment applications, we are also collaborating with Mitsubishi Electric to develop an artificial intelligence guidance system for both precaution warning and set-point guidance at Bedok NEWater Factory (BNF).

The aim of the system is to warn operators in advance of impending process anomalies, and to recommend the relevant control strategies (i.e. set-points) for process troubleshooting. This helps to bridge the "knowledge gap" that new operators might have, while they are familiarising themselves with the plant processes.

A smart Supervisory Control and Data Acquisition (SCADA) system to be developed by Surbana Jurong also studies interactions between various parameters at the Choa Chu Kang Waterworks to anticipate process anomalies and pinpoint their causes.

Data analytics and machine learning could also transform leak detection processes. Currently, PUB verifies leaks in its potable water network by flagging anomalies in pressure transient and acoustics data collected by sensors in its WaterWiSe system, and then manually combing the vicinity of the anomalies.

This approach, however, generates a high rate of false alarms. In order to zero-in on the leaks sooner, we have partnered with software firm Bentley Systems to develop a solution to single out anomalies and localise them in near real-time using machine learning.

MIRROR UNIVERSES

Digital twins are the third major group of process and system optimisation enablers. Utilities can use virtual copies of plants to simulate the consequences of various actions, whether to better guard against cyberattacks, ascertain the best option in an array of them, or train operators to prepare them for events such as power or equipment failures.

Apart from the PUB-Yokogawa project covered in the previous chapter, PUB has put in place a whole plant simulation platform co-developed with the Jacobs Engineering Group in the CWRP, and another one with Royal HaskoningDHV in the IVP.

The Jacobs platform replicates CWRP's processes, including its process, hydraulic and control systems, and liquid and solid streams. It retrieves real-time production and process data from the control and monitoring system, analyses the information to strip out errors, and feeds it into the digital twin to study the plant's performance and make predictions.

By comparing its predictions of the plant's performance to the plant's actual performance, it can alert operators to irregularities that need their attention. It can also look up to five days ahead, estimate the likelihood of specified events, warn operators of imminent deviations in treatment processes and poor effluent quality, and recommend steps for rectification.

These capabilities have enhanced the resilience of operations at the CWRP. As the simulator runs independently of the plant's actual systems, it can also be used for operator training, and to safely evaluate operating strategies before these are put into practice.

For example, if the operators have to lower the solids retention time in the bioreactor to take some volume out of service, they could use the simulator to determine the impact on the quality of the secondary effluent, and how increased waste activated sludge production could affect the thickening and digestion processes.

THE ROAD AHEAD

As we continue on our digitalisation journey, there are still many avenues that are ripe for exploration. With the data from the upcoming smart water meters network and more detailed rain forecasts, we could better project sewer flows and the intake of water reclamation plants. Data from other agencies and even other water utilities globally could also be helpful.

To make the most of digitalisation, utilities need to ensure that they are collecting relevant and good quality data, and have data scientists and their domain experts working hand-in-hand.

When creating and trialling systems, utilities should deploy them when the accuracy reaches 80%, and then continue to refine them based on feedback. No system, especially machine learning ones, will be perfect from the beginning. In digitalisation, as in many other endeavours, the most important thing is to start.

Chapter 5

Cybersecurity: the backbone of digitalisation

As PUB embraces digitalisation, we have also continued to make cybersecurity a cornerstone of our policies, plans and projects. While cybersecurity has always been a critical priority for us, our increasing degree of digitalisation has necessitated more expansive and stringent protections, cybersecurity training and failsafe measures.

Across the globe, the risk of cybersecurity attacks and breaches has risen in tandem with digitalisation, due to the interconnectedness of devices and physical assets. In 2019, a report by technology firm Siemens and data security research centre the Ponemon Institute found that cyber threats to both water and electric utilities are becoming more frequent, severe and sophisticated.

About half of the 1726 utility respondents said that their organisation had experienced a cyberattack in the past year that led to the loss of private information or outages in operational technologies. About half of the respondents expected another attack in the next year.

Malicious attacks can impact water utilities in an array of ways. In Australia, a utility was infiltrated by a disgruntled former contractor who used his access to release nearly one million litres of raw sewage into rivers, parks and residential grounds.

© IWA Publishing 2020. Digitalising Water – Sharing Singapore's Experience
Author: PUB, Singapore's National Water Agency
doi: 10.2166/9781789061871_0047

In the United States, a hacker stole 2.5 million records from a utility by taking advantage of a wired connection between an Internet-facing payment application server and a computer system that served operational and info-communications functions.

In Britain, a utility discovered that an employee in its third-party call centre which administered online accounts and processed telephone payments was bribed to redirect more than £500,000 in refunds to overseas bank accounts.

The list goes on. To protect themselves, water utilities must make cybersecurity a priority. Israel recently successfully thwarted a major cyberattack against its water systems in April 2020, which was a synchronised and organised attempt at disrupting key national infrastructure.

PUB has invested in its comprehensive cybersecurity policy, and put in place systems, processes and training to enforce it.

SYSTEMS AND DESIGNS

From the beginning, our infrastructure is designed and built to resist cybersecurity attacks. PUB has a Water Sector Policy that includes provisions from Singapore's Cybersecurity Act and best practices from the United States' National Institute of Standards and Technologies. These guidelines ensure a baseline level of protection for its assets.

As a matter of policy, for instance, systems are not connected to the Internet or cloud technologies to limit their exposure to malicious actors.

When each plant is being designed, it has to conform to a cybersecurity design framework so that cybersecurity measures and allowances are embedded from the start. This includes conducting threat assessments for systems to determine their vulnerabilities and address any security gaps before the systems are installed.

Within installations, biometric systems restrict access to control rooms and server rooms, incorporating multi-factor authentication mechanisms for critical zones.

Internal control systems are air-gapped and allow only the unilateral flow of information to external networks, preventing hackers from tampering with the plant controls to cause harm or catastrophic consequences.

Apart from these and other protective measures, detection systems are put in place to alert operators to potential intrusions and breaches, such as valves that switch on or off without their instructions, and attempts to export data.

Such forethought is crucial as retrofitting plants and systems to boost their cybersecurity is much more difficult and may require processes to be temporarily shut down, affecting operations.

As an added precaution, we also conduct penetration testing when there are major changes to systems, such as software modifications, as well as annually, to make sure that the systems remain robust against cyberattacks.

TRAINING AND PROTOCOLS

All our staff also undergo cybersecurity training to inculcate basic safety principles and practices. These include using only secured thumb-drives to transfer data, as external ones may contain malware, and not clicking on emails from unknown senders, as these may be vehicles for phishing attempts and other forms of cyber intrusions.

Our shift leaders who supervise operators are taught to recognise signs of cybersecurity incidents, such as unexpected changes in pump operations or water quality readings. While hackers may attempt to hide their tracks by manipulating sensors related to the operational processes they are trying to hijack, the data derived from other sensors may expose them.

Our Instrumentation and Control (IC) engineers who are in charge of Supervisory Control And Data Acquisition (SCADA) systems as well as operational and maintenance ones receive even more training, including a five-day programme that covers cybersecurity defences and what to do during attacks or breaches.

PUB has also assembled a cyber emergency response team that consists of cybersecurity professionals and specialists. This apex team steps in when a cybersecurity incident is unable to be solved by the operators, shift leaders and IC engineers.

BACK-UPS AND RECOVERY

Despite our best efforts, some cyberattacks may succeed. To minimise the consequences, we have also invested in recovery measures.

Our industrial control systems have redundancies, where in the event if one system or component is shut down, there is a backup system or component in place to ensure that operations function as normal.

Moreover, data and system configurations are backed-up regularly and saved offline so that these can be deployed for continuity of operations, reducing the time needed for recovery.

SECURITY FIT FOR FUTURE

With cybersecurity attacks and defences constantly evolving to battle each other, we have also taken steps to ensure that our protection systems remain up to the mark.

In 2015 and 2016, we worked with iTrust, the Singapore University of Technology and Design's centre for cybersecurity research, to establish the Secure Water Treatment (SWaT) and Water Distribution (WADI) testbed facilities.

The SWaT testbed replicates a six-stage process for treating water. The WADI testbed, for its part, comprises reservoir tanks, consumer tanks, raw water tanks, a returned tank, chemical dosing systems, booster pumps and valves, instrumentation and analysers. It also takes in a portion of the SWaT testbed's reverse osmosis permeate.

Together, the two testbeds form a complete and realistic water treatment, storage and distribution network, providing researchers across the globe with an industrial-grade and one-of-a-kind platform to conduct experiments and research to validate models and theories about cyberattacks, and to test the robustness of cyber protection methods.

After years of intense research, several patents have been filed for novel technologies aimed at detecting cyberattacks from multiple angles and enabling systems to recover from them quickly. These include a system that creates unique "noise fingerprints" for sensors, to detect deviations that are caused by sensor spoofing attacks.

As cyberattacks become more innovative, cybersecurity measures need to follow suit. The SWaT and WADI testbeds are essential in helping PUB and other water utilities to stay ahead of cyber threats.

Beyond a commitment to cybersecurity research, we will also continue to upgrade the cybersecurity competency of our staff. For instance, we have partnered with the Civil Service College Singapore to craft a competency framework for officers whose primary work is in cybersecurity, to guide them in building up and updating their store of knowledge and skills.

Such professional development is also essential as there is a shortage of capable cybersecurity personnel, who are in high demand not just in the water sector but in many other industries. This is partly because cybersecurity has traditionally focused on information technology rather than operational technology.

For this reason, it is often easier to train plant engineers in cybersecurity. Their domain expertise in water utility operations is also advantageous when it comes to designing secure systems for use in plants.

Looking ahead, we will also enhance our security operations centre to boost cybersecurity. The centre currently follows up on security alerts that are triggered in the water network. In the future, we will harness machine learning technologies to expand our capabilities, allowing better detection of anomalies that may indicate cybersecurity attacks and breaches.

We will also gather intelligence on cybersecurity attacks, espionage, malware and other threats globally that have befallen water utilities or may potentially endanger them. Such analyses will serve as a leading indicator for PUB's cybersecurity risks.

Networking and sharing of information with other organisations are also priorities. PUB already shares information and best practices with other utilities and agencies, and gleans the same from them, through study trips, events, conferences and meetings such as the annual Singapore International Cyber Week.

We are also a member of Singapore's Operational Technology Information Sharing and Analysis Centre, which is a threat information sharing hub for companies and organisations in the energy, water, and other critical information infrastructure sectors.

Members can securely exchange details of operational technology and information technology threats and attacks on their organisations, to prevent and quickly mitigate and contain damage caused by malicious actors.

The threat of cybersecurity attacks and breaches should not deter water utilities from embracing digitalisation. By taking a proactive approach to defence, they can minimise their vulnerabilities, and forge ahead to seize opportunities.

Chapter 6

The human element in digitalisation

Ultimately, the success of our digital transformation rests on our employees. Although digitalisation is about making use of hardware and software, it can only succeed if it is supported by people, both in spirit and in execution. A water utility can only transform itself into a smart utility of the future if the change is led by strong leadership and buttressed by employees at all levels.

In PUB, the push for digitalisation starts at the top. The utility has established a Smart PUB Steering Committee chaired by the Deputy Chief Executive of Policy and Development and Deputy Chief Executive of Operations. The committee's work is also coordinated by PUB's Chief Information Officer.

To inspire employees, PUB's senior leadership has also highlighted the vision of a digitalised utility in speeches and townhall meetings, where work is simplified, and made easier and more efficient through automation and other technologies, where employees can achieve better work-life balance.

In 2019 alone, PUB's senior management hosted over 200 town hall meetings with over 3,000 officers across the organisation. Through these gatherings, they shared the impetus for the utility's digitalisation drive, and emphasised that all officers play a role in the endeavour.

The senior leaders also helm key transformation projects throughout the utility, such as the upgrading of the Intelligent Water Management

© IWA Publishing 2020. Digitalising Water – Sharing Singapore's Experience
Author: PUB, Singapore's National Water Agency
doi: 10.2166/9781789061871_0053

System, leveraging their presence to boost support and momentum for change.

A UTILITY-WIDE EFFORT

To encourage buy-in from the rank-and-file, all internal communications and publicity for the digitalisation movement are designed to showcase our transformation vision, the reasons for the need to change, how officers can participate, the rewards for participating, and how digitalisation can and must take place across all levels of the utility.

PUB also has a one-stop innovation portal for all officers called the PUB IDEAS Exchange. In this portal, our officers can view, learn from and contribute to all ideas, including proposed and implemented ones, and prototypes created by their peers.

For every realised idea with cost savings, up to S$15,000 is given to the officers behind it, to motivate all to continue innovating at work.

We also recognise the efforts of our officers through internal newsletters and features, and an annual Service and Innovation Day (SID) to recognise the best innovation and service staff in the utility. Outstanding ideas in the past year also qualify for the Chief Executive Innovation Awards for additional monetary awards.

The SID is particularly important as a platform for staff to relate their personal stories on innovation and digitalisation, which in turn inspires and motivates others to join the utility's journey. These stories often have common themes of determination and teamwork, which reinforce the values that are needed for success.

The SID is also accompanied by an Innovation Exhibition featuring the latest innovations and technologies that are being incorporated in PUB. Having such an exhibition demonstrates to the utility's officers that change is real, and is happening concurrently across the breadth of operations, to excite them about the benefits that they will experience in the coming months.

BASIC TRAINING FOR SUCCESS

While enthusiasm for digitalisation is essential, so is having the relevant knowledge and capabilities. At PUB, providing employees with digitalisation training tailored to their job scope and responsibilities is a priority.

In 2018, the Singapore government set the goal of training all public service officers to have basic digital literacy by 2023, and basic competency in digital skills. The Public Service Division (PSD) set out a Digital Literacy Framework that outlines the basic standards required of all public officers to work in a digitally-enabled workplace.

Guided by PSD's framework, PUB's Singapore Water Academy (SgWA), which offers training for water professionals locally and internationally, formulated a digital literacy training approach contextualised to the utility's needs. It has three tiers of standards that specify the digital knowledge and skills required by different job grades.

The academy has sourced and crafted courses to train officers in these skills. These include Digital Workplace courses that cover technology and data literacy, data protection, cybersecurity, technologies shaping digital transformations, applications of data analytics, hands-on training in data collection, and much more.

As of 2019, about 75% of the PUB officers selected for the courses have completed them, and we are on track to finish the training by 2021.

As change management is a critical skillset for senior leadership, project leaders and other change agents, the SgWA has also rolled out courses in this field.

ZEROING IN ON DATA ANALYTICS TRAINING

Building a future-ready data analytics team is also necessary for water utilities to make the digitalisation leap. Many organisations are only 'data-aware', where their employees simply make spreadsheets and compile them for reporting. The goal is to be data-driven, where decisions are made through data, where machine learning and forecasting models are heavily utilised.

The SgWA, together with the InfoTech and Digital Transformation Department, has contextualised the Singapore Government Technology Agency's Data Science Competency Framework to introduce a data analytics training framework aimed at giving staff the relevant knowledge and skills to make data-driven decisions. This framework has several proficiency tiers that are mapped to PUB's job roles based on competency requirements.

The plan is to create a large base of officers with basic proficiency in data analytics, and spur officers with interest and motivation to attend data analytics courses to further deepen their skills and knowledge.

PUB also plans to train a group of data analysts to be highly proficient in business intelligence analytics, and to produce data engineers who can design and actualise big data architectures.

To accomplish these objectives, we will organise basic data analytics training for all our engineers and executives, and place data analysts in various departments in intermediate and advanced data analytics programmes.

With a strong data analytics culture within the organisation, our staff will cultivate the habit of using data and its insights to support decisions, and thus improve the utility's performance.

SETTING SIGHT ON DESIGN INNOVATION

We are also implementing training to equip our officers for design innovation. Such innovation involves examining and rethinking work processes to increase productivity and achieve other work targets. Data analytics is indispensable to gather evidence and validate assumptions during such exercises.

PUB's InfoTech and Digital Transformation Department has partnered with the Singapore University of Technology and Design (SUTD) and SgWA to curate a six-day course to impart officers with the design innovation and data analytics skills required to facilitate work process reviews and their redesign.

The course fosters a collaborative approach involving multidisciplinary teams working within the organisation to come up with fresh perspectives to improve its products and business processes.

It also includes a data analytics bootcamp to instruct officers in the subject. After finishing the course, participants will have foundational knowledge of data analytics and be able to apply design innovation tools while taking charge of transformation projects in their departments.

The inaugural course was conducted in November 2019 at SUTD for 37 officers.

VALUING PEOPLE

Going forward, we will also devise more opportunities for PUB employees to work alongside contractors in the development and implementation of digitalisation solutions, to give them more hands-on experience in the field.

PUB Chief Executive Ng Joo Hee stressed that the utility's people are its most valuable assets.

He said: "We have to maximise the full potential of our people, and enable them to lead and to innovate for extraordinary results. We have to give them the best training and development throughout their time with PUB in order that they can be as high performing as possible. Organisational change doesn't happen by itself. It needs to draw strength from a common vision. It needs to be sustained by a pervasive culture that embraces change. The need to change, how to change, what to change, and what to change next has to be communicated up and down our organisation, such that even the lowliest operator in the farthest outpost knows that he too has to do things differently."

Chapter 7

Conclusion

When a football team is down by one man, the remaining players need to play smarter and adjust their tactics to win. That is the analogy that PUB Chief Executive Officer Ng Joo Hee uses to explain the importance of PUB's push for digitalisation.

In the near future, some utilities may need to learn to work with fewer employees due to ageing populations and low birth-rates. Other countries may face different forms of resource restrictions, such as a growing lack of rainfall for harvesting, treatment and use due to climate change. Digitalisation can offer solutions for these and other issues.

In this publication, we have summarised PUB's key learning points from its digitalisation journey so far. Building and choosing the right infrastructure is paramount, and this refers to not just hardware such as wireless communications networks and sensors. Software systems to break down information silos, and to aggregate, supplement and analyse data, are crucial too.

As utilities construct their infrastructure, they must also make cybersecurity a cornerstone of their plans. This means ensuring that cybersecurity measures are considered and built into plant design from the very beginning, and additional technologies are vigorously tested before they are implemented in operational systems.

© IWA Publishing 2020. Digitalising Water – Sharing Singapore's Experience
Author: PUB, Singapore's National Water Agency
doi: 10.2166/9781789061871_0059

Moreover, manpower is the difference between failure and success. PUB has appointed a Chief Risk Officer and a Chief Information Officer to helm new departments leading its digital transformation to become a smart utility of the future. Its variety of training programmes will equip staff with the skills necessary for this transformation.

The projects featured in this publication also underline the benefits of collaboration. In PUB, research and development and industry development have been brought together under one Assistant Chief Executive, to bring discoveries from lab to application as quickly as possible.

By forging partnerships, utilities can also mine the rich lode of expertise and resources in firms and research organisations, and in turn offer real-world testing facilities to speed up the commercialisation of innovations in water to the marketplace. As the saying goes, many hands make light work.

DIGITALISATION TAKES CENTRE-STAGE

Many water utilities are on the path to digitalisation, signalling the growing recognition of its potential.

Umgeni Water, a utility in Durban, South Africa, deployed digitalisation to improve its water resource management and better protect customers. This included pairing hydrologic models and monitoring devices to optimise storage levels in dams and reservoirs.

The Las Vegas Valley Water District, the government agency that supplies the Las Vegas Valley with water, turned to digitalisation to locate leaks. It uses leak detection devices to listen for sounds and vibrations caused by water seeping from its distribution lines.

By monitoring the pipes continuously and fixing small leaks before they become larger, the agency can prevent major water supply disruptions, and has extended the lifespan of the pipes.

Anglian Water, which provides water and water recycling services to more than six million people in the United Kingdom, is partnering with engineering firm Black & Veatch to create a digital twin of the utility's water treatment and distribution infrastructure in the Newmarket region.

Artificial intelligence (AI) with predictive and analytic capabilities will be embedded into the digital twin to support the utility's decision-making in the region.

In Malaysia, Syarikat Air Melaka Berhad (SAMB), the state of Melaka's water supply company, is increasingly relying on digitalisation too. It has installed a Geographic Information System (GIS) that contains consolidated and regularly-updated data about its pipe network, including pipe sizes, materials and connections.

When customers call to report pipe malfunctions, an automated locating system allows SAMB to alert the nearest repair team to reach the site within 20 minutes. The team checks the problematic pipe's details through the GIS on-site and orders replacement parts immediately. This has reduced supply downtime from up to 24 hours previously to 3-4 hours currently.

Like PUB, these and many other utilities have made digitalisation a core component of their operations, and an essential part of their future.

AT THE FOREFRONT OF DIGITALISATION

As our digital evolution continues, we are also keeping an eye out for promising nascent digital technologies, and ways that these could be adapted for Singapore's current and future needs.

Blockchain, for instance, is beginning to be trialled in several utilities worldwide. In California in the United States, tech giant IBM is working with non-profit organisations, other firms and researchers to test the use of blockchain, sensors and other Internet-of-Things technologies to monitor groundwater usage transparently and in real-time.

The sensors transmit water extraction data which is stored by IBM's blockchain platform to serve as an append-only, immutable record. A web-based dashboard available to farmers, financiers and regulators enables all parties to track the groundwater usage, and individual users to trade their allotted usage.

AI chatbots can also answer water users' questions quickly so they do not have to be on hold with a call centre or search through pages of information on their utility's website. Northern Ireland Water's AI chatbot handles 2,500 to 3,000 customer interactions monthly, recognises customers' questions 92% of the time and answers correctly 70% of the time.

As digital solutions become more sophisticated and new ones enter the market, PUB will evaluate them and possibly even help to develop several of them, as we have done in the past and continue to do today.

PUB's Mr Ng noted that the utility's digitalisation plan will propel it wholly and firmly into the world of digital technology. He said: "There are many changes that are underway throughout the length and breadth of our organisation, and we have every intention to press ahead and persevere with these changes until they become the new normal in PUB."

He summarised: "We cannot carry on as usual. We have to change, and we have to change now. Our water mission is too important."

IWA Publishing's authorised EU representative for General Product Safety Regulations is Diane D'Arras, 15 rue Duret, 75116 Paris, France, e-mail: safety@iwap.co.uk.

Printed and bound by CPI Group (UK) Ltd, Croydon, CR0 4YY

13/05/2026

02109710-0001